BUG HUNT

Storey Publishing

Are You Ready to

SEARCH FOR BUGS?

The world is full of all kinds of cool bugs. You can find them buzzing, flying, climbing, and crawling nearly anywhere if you pay attention! Grab this book, head outside, and get ready to spot some little critters.

THINGS TO BRING IN YOUR BACKPACK

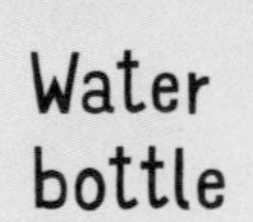

Water bottle

Camera

Bug jar

This book and magnifying glass

Pencil

Sun hat

Bug spray

Snacks

Sunscreen

Bug net

HOW TO HAVE FUN WITH THIS BOOK

Use the MAGNIFYING GLASS

Weave SPIDERWEBS

Listen for CHIRPING and BUZZING

Search for JUMPING, FLYING, and CRAWLING CRITTERS

Look for BUG SIGNS

POLLINATE flowers

Play BUG GAMES

Meet your NEIGHBORHOOD INSECTS

BUG PATCH STICKERS

There are 12 patch stickers in the back of this book that match **I SEE IT!** circles on some of the pages. When you see a bug that matches one on an **I SEE IT!** page, put the sticker on the matching circle. See how many you can find!

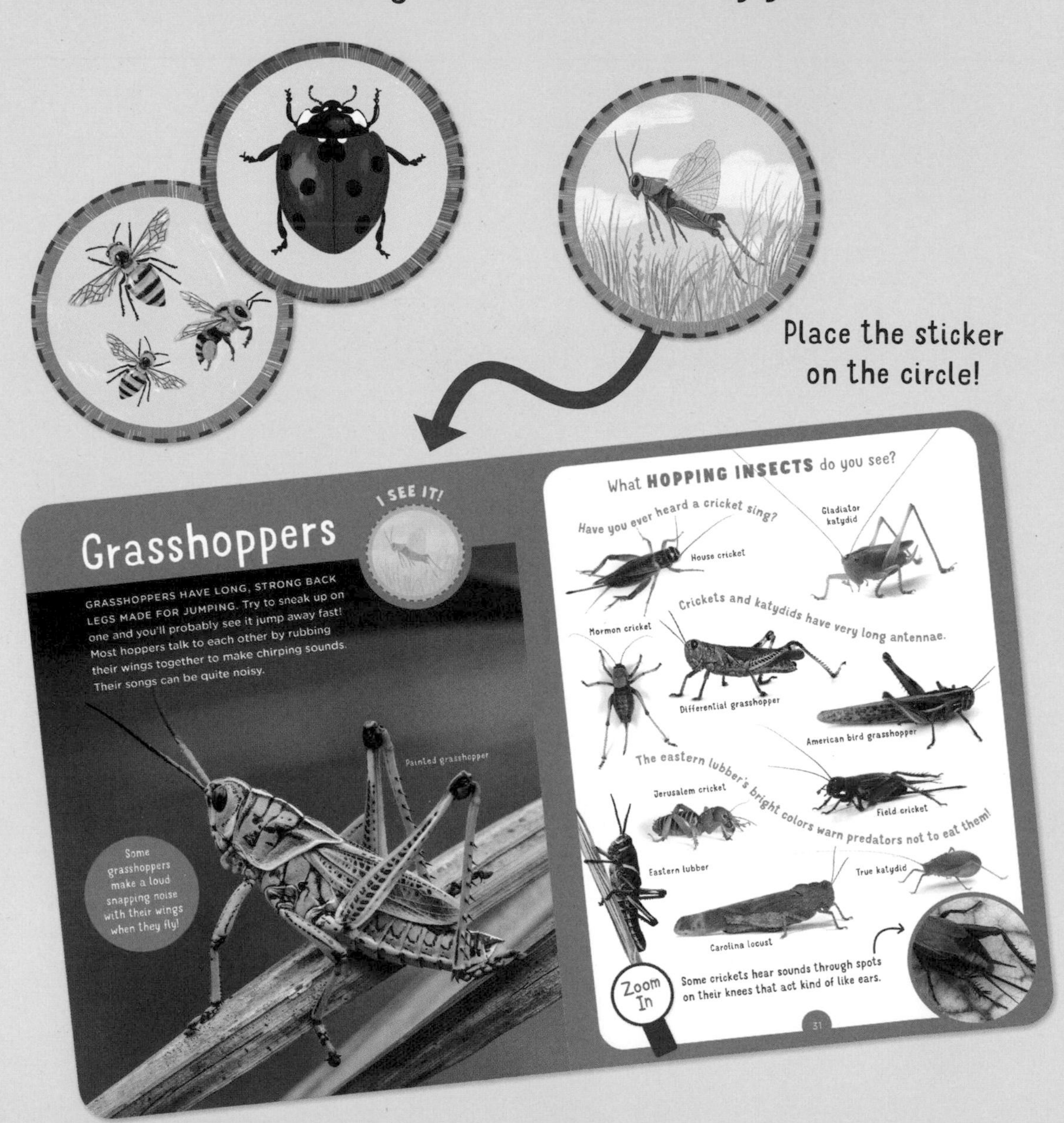

Place the sticker on the circle!

I SEE IT!

Grasshoppers

GRASSHOPPERS HAVE LONG, STRONG BACK LEGS MADE FOR JUMPING. Try to sneak up on one and you'll probably see it jump away fast! Most hoppers talk to each other by rubbing their wings together to make chirping sounds. Their songs can be quite noisy.

Painted grasshopper

Some grasshoppers make a loud snapping noise with their wings when they fly!

What **HOPPING INSECTS** do you see?

Have you ever heard a cricket sing?

House cricket

Gladiator katydid

Crickets and katydids have very long antennae.

Mormon cricket

Differential grasshopper

American bird grasshopper

The eastern lubber's bright colors warn predators not to eat them!

Jerusalem cricket

Field cricket

Eastern lubber

True katydid

Carolina locust

Zoom In

Some crickets hear sounds through spots on their knees that act kind of like ears.

31

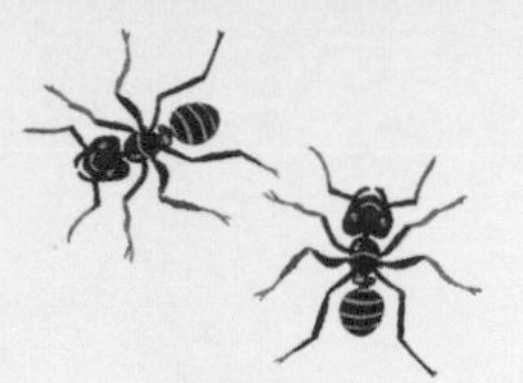

Bug-finding TIPS

Here are some ways to find bugs and learn about them.

LOOK IN THE AIR AND ON THE GROUND.
Try peeking in flowers, checking under logs, and looking in shallow water.

BE GENTLE.
Don't harm or squish creatures.

NOTICE THE SHAPE, SIZE, AND COLOR of a bug to help you identify it.

LISTEN QUIETLY.
Your ears can help you find insects that make sounds.

BE CAREFUL AROUND INSECTS THAT MIGHT BITE OR STING.
Give them space.

Caterpillars

I SEE IT!

CATERPILLARS ARE BABY MOTHS AND BUTTERFLIES. They spend most of their time eating leaves and growing. After a while, moth caterpillars spin a cocoon and butterfly caterpillars make a chrysalis [KRI-sah-lis]. Then they grow their wings and turn into adults.

Tent caterpillars hang out together in silky webs.

What kinds of **CATERPILLARS** can you find?

This caterpillar is known as a woolly bear.

Isabella tiger moth

Luna moth

Monarch butterfly

This caterpillar is making its chrysalis.

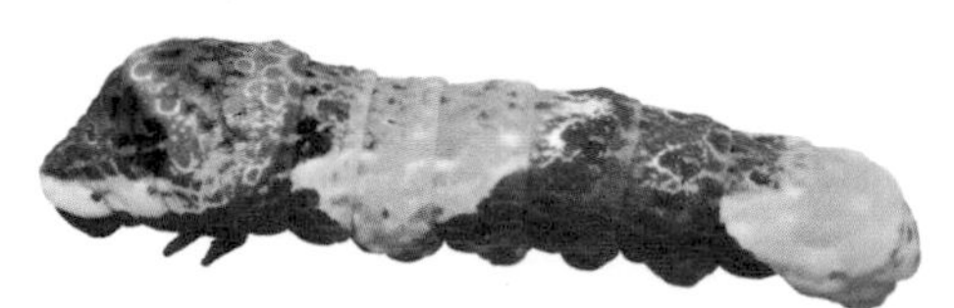

Giant swallowtail butterfly

Caterpillars come in lots of colors.

Tobacco hornworm moth

Milkweed tiger moth

Mourning cloak butterfly

Some caterpillars have stinging spines to keep predators away.

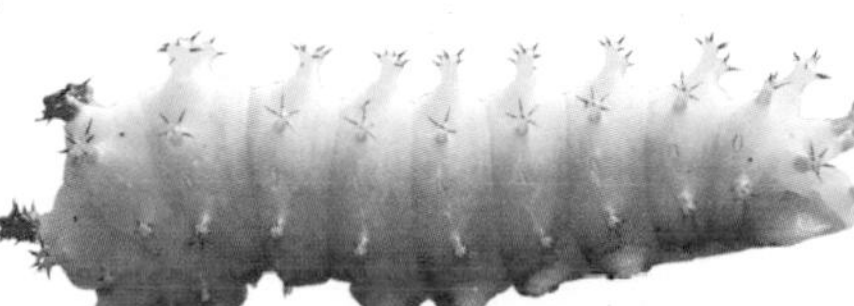

Cecropia moth

Is that crawly caterpillar smooth or furry? What color is it? Does it have stripes? Spots?

WATCH A CATERPILLAR GROW!

Metamorphosis [meh-tuh-MAWR-fuh-sihs] is when something changes from one form into another.

Follow the pictures below to see the life cycle of a monarch as it changes from egg to caterpillar to chrysalis to butterfly!

1. A female monarch lays her tiny **EGGS** on the underside of a milkweed leaf.

2. Between 4 and 8 days later, a little caterpillar wiggles out of each egg. The caterpillar is a **LARVA** that eats a lot for 2 or 3 weeks. It grows and grows!

3. When the caterpillar is big enough, it hangs upside-down from a twig and forms a **CHRYSALIS**.

4. In about 10 days, it emerges as a **BUTTERFLY**!

BUG BEHAVIOR

Insects are always on the move. If you notice a bug behaving in any of these ways, check it off the list!

Butterflies

I SEE IT!

THESE GENTLE INSECTS START THEIR LIVES as hungry caterpillars before they transform into beautiful butterflies. Some caterpillars turn into moths. Butterflies have smooth antennae and visit flowers during the day. Moths look like butterflies but have feathery antennae. They are often nocturnal [nok-TUR-nul], which means they fly mostly at night.

Butteflies and moths have long, tubelike mouths to suck up flower nectar. The tube is called a **PROBOSCIS** [pruh-BAH-skihs].

Monarch butterfly

What kinds of **BUTTERFLIES** and **MOTHS** do you see?

Butterfly and moth wings are covered in tiny, colorful scales.

Buckeye butterfly

Viceroy butterfly

Spicebush swallowtail butterfly

Butterflies and moths are found everywhere except in Antarctica.

Cloudless sulphur butterfly

White-lined sphinx moth

Butterflies and moths have two pairs of wings.

What's your favorite butterfly?

Mourning cloak butterfly

Hummingbird moth

Green hairstreak butterfly

Painted lady butterfly

Butterflies and moths taste with their feet!

Zoom In

See how a moth's antennae look like little feathers?

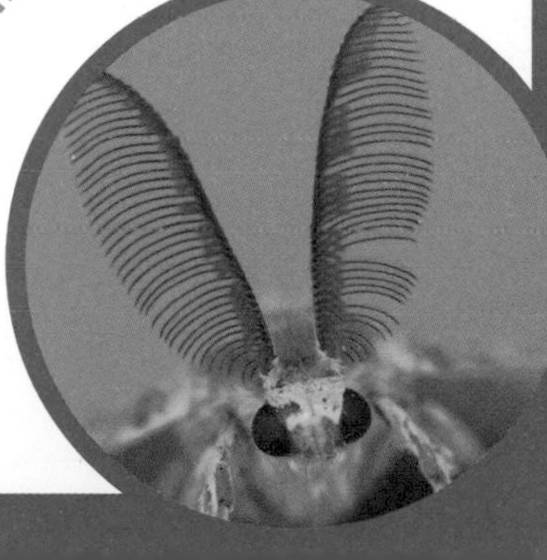

WHAT iS iT?

Some people call any crawly critter they see a "bug," but the scientific name is arthropod [AHR-thruh-pod]. Arthropods have a hard outer shell instead of a skeleton. They come in all shapes, sizes, and colors, but here are some easy ways to tell different kinds of arthropods apart.

There are more arthropods on Earth than any other type of animal.

Is iT AN iNSECT?

Most insects have six legs and one pair of antennae. They have three main body parts — head, thorax, and abdomen. Most also have wings. Some insects munch plants or sip nectar. Some insects eat other insects!

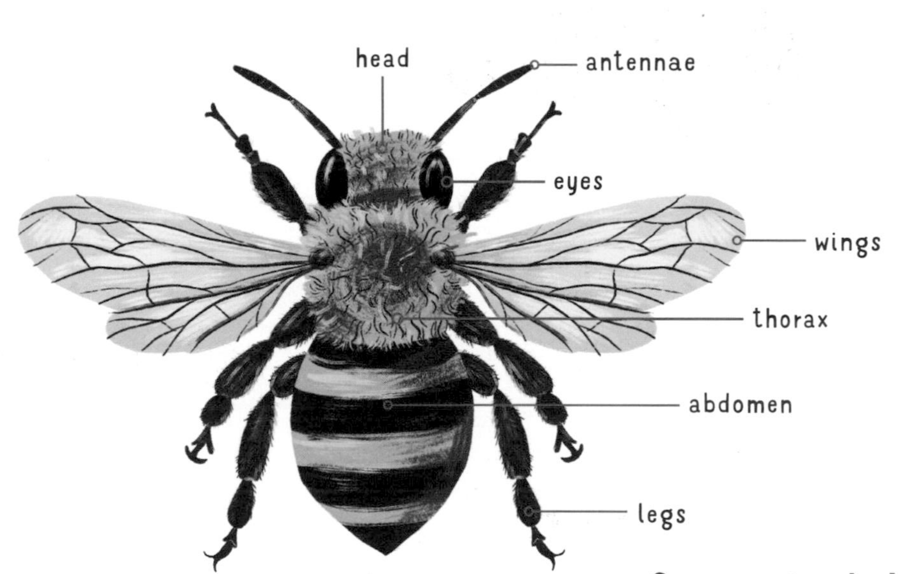

Bumblebee

Some animals have bones on their insides, but arthropods have a hard outer shell called an exoskeleton. It protects them like a suit of armor.

Is it a bug?

Not all insects are bugs, but all bugs are insects. True bugs have mouths like straws for sucking up their food.

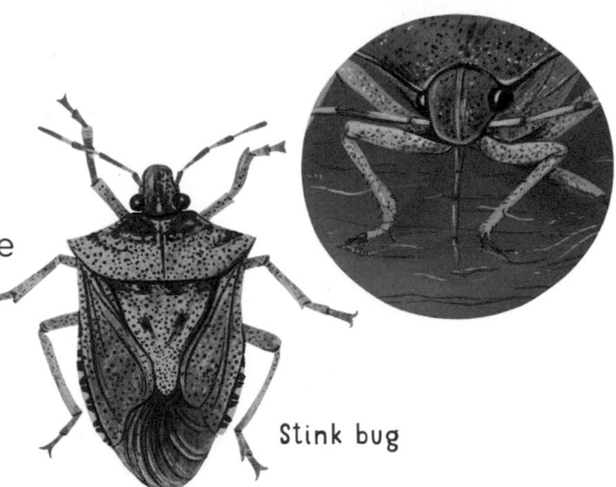
Stink bug

Is it a spider?

Spiders, scorpions, and ticks are not insects. They have eight legs, and they do not have antennae or wings. They have two main body parts — cephalothorax [SEH-feh-leh-THOR-ax] and abdomen. Most hunt other animals for food.

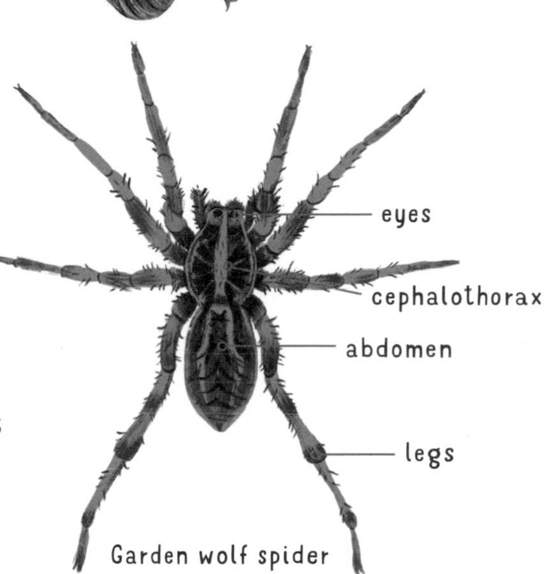

Garden wolf spider

Millipede

Centipede

Millipedes and centipedes belong to a group called myriapods [MIHR-ee-uh-pods], which means "many legs." They can have up to 700 legs!

Centipedes have one pair of legs per body segment, while millipedes have two pairs per segment.

Bees

I SEE IT!

THOSE YELLOW AND BLACK STRIPES WARN OTHER ANIMALS TO STAY AWAY if they don't want to be stung. Some kinds of bees live alone, but others live together in groups called colonies. All bees drink flower nectar but only honeybees use that nectar to make sweet honey in their hives.

Carpenter bee

The buzzing noise a bee makes comes from its beating wings!

What kinds of BEES and WASPS do you see?

Yellow jacket

European hornet

Paper wasp

Metallic green sweat bee

Bumblebee

Bald-faced hornet

European honeybee

Black-and-yellow mud dauber

Zoom In

If you see a bee in a flower, look for golden or orange pollen on its legs or tummy.

POLLINATE A FLOWER!

Bees and other insects help grow our food. When a bee crawls into a flower looking for sweet nectar, it picks up sticky pollen on its legs.

When it flies to a different flower, some of that pollen rubs off inside. That's called pollination [pol-luh-NAY-shuhn].

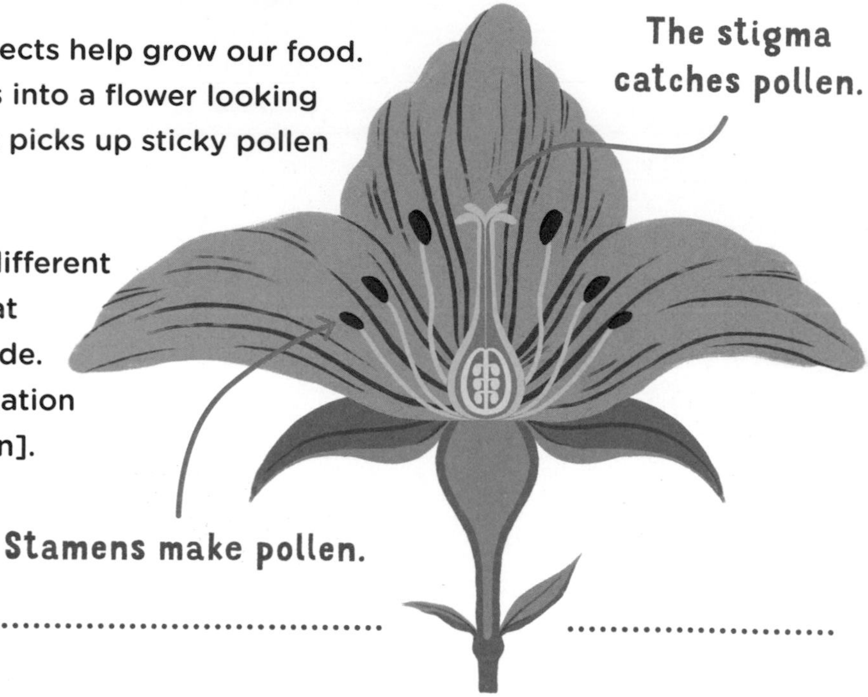

The stigma catches pollen.

Stamens make pollen.

YOU TRY IT!

Follow these steps to help a plant make a seed that will grow into another plant!

1. Find a small paintbrush, a long piece of grass, or a skinny twig. If you don't mind getting your hands dirty, a finger works too!

2. Look for a patch of blooming flowers. Look inside a blossom to find the plant's dusty orange or yellow pollen.

3. Gently touch the pollen with your paintbrush, grass, twig, or finger until some rubs off.

4. Now find another flower that is the same type. Rub the pollen on the flower's long, sticky stigma.

Plants need pollination to make seeds, so pollinating insects are really important.
Capture some pollen from one flower.
Place the pollen on another flower.

Ants

I SEE IT!

ANTS CAN'T SEE VERY WELL, so they learn about the world around them by smelling and touching. They use their long, thin antennae to test food and sniff their way back home. When two ants meet on a trail, they feel each other with those antennae to find out if they are friends or enemies.

Leaf-cutting ant

Ants can carry loads much heavier than themselves.

What kinds of **ANTS** do you see?

Watch a colony of ants crawling in and out of their underground nest.

MAKE A BUG HOTEL

With a simple bug observation jar you can watch your favorite bugs for a short time.

YOU'LL NEED:

- A clean, clear plastic container or sturdy glass jar
- A small square of cloth or paper towel
- A rubber band

1. When you see a spider, beetle, or caterpillar you'd like to study, gently scoop it up and put it in your jar.
2. Look at the kind of plants your critter was crawling on. Collect the same type of leaves and small twigs and add these to your jar.
3. Place the cloth or paper towel over the top of the jar and hold it in place with the rubber band. This lets air enter the jar while keeping your friend inside.
4. Sit and watch what your insect is up to!
5. After you've watched your bug for a while, let it go where you found it.

Ask questions while you observe your insect.

How does it move around?

What color is it?

What type of bug is it?

Is it eating leaves?

Ladybugs

LOOK FOR LADYBUGS IN THE GRASS OR ON PLANTS. If you find one, count its spots! Like most beetles, ladybugs have two pairs of wings. The tough spotted front wings protect the soft inner wings they use to fly. Beetles' outer wings come in many beautiful colors and are often bright and shiny.

Gardeners love ladybugs because they hunt plant-eating bugs like **APHIDS** [AY-fihds].

What other **BEETLES** can you find?

Big Dipper firefly

Giant stag beetle

Eyed click beetle

Green June bug

Eastern Hercules beetle

Ten-lined June beetle

Potato beetle

Rainbow scarab dung beetle

Click beetle larva

Zoom In

A beetle's hard "back" is really a pair of stiff outer wings.

GLOW IN THE DARK

Fireflies have a special way of finding each other in the dark. They glow! These flying beetles make a special chemical in their abdomens that they flash on and off. Their larvae glow too, which tells predators they taste bad.

If you live in an area with fireflies, you can catch them in a jar to make a lantern. Always let them go after a little while.

It takes about 70,000 fireflies to create the same amount of light as one lightbulb!

Look for fireflies on warm, humid summer nights in the eastern half of the United States. How many can you count?

BUG SIGNS

Don't see any bugs around you? Keep on the lookout for these bug signs. If you see any, check them off here.

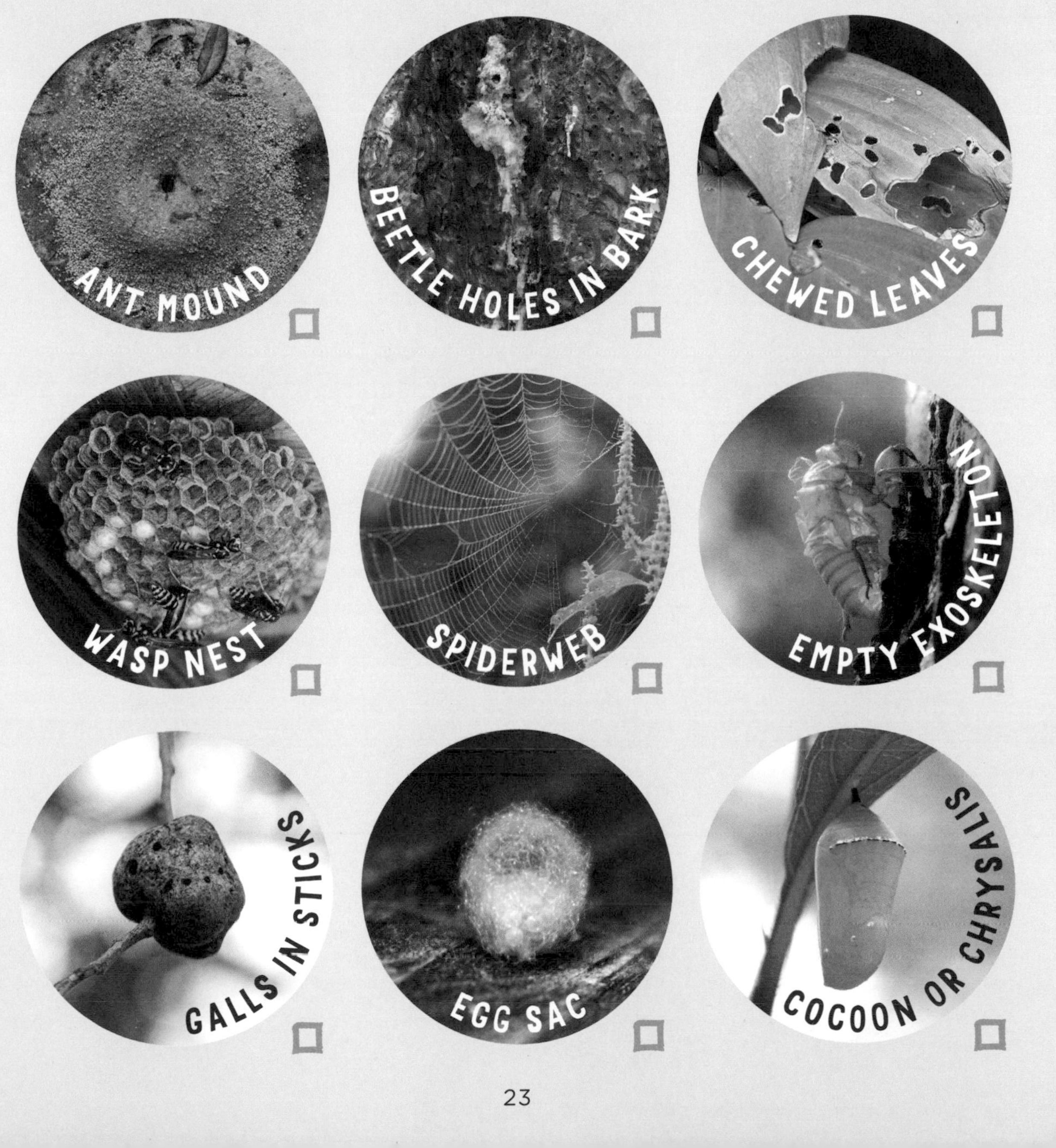

Dragonflies

I SEE IT!

DRAGONFLIES ARE AMAZING FLIERS. You may see them zooming through the air by a lake or pond, chasing each other, or hunting for smaller bugs to eat. They are really fast and can even fly sideways and backward! Their long, thin wings are see-through like window glass.

Familiar bluet damselfly

Damselflies are slender dragonflies. Their long, skinny bodies help them balance while flying.

Can you spot a DRAGONFLY zooming around?

Prehistoric dragonflies were the size of seagulls!
Familiar bluet damselfly
Eastern pondhawk
Blue-tailed damselfly
Twelve-spot skimmer
Dragonfly larvae live underwater before becoming adults.
Blue dasher
Red saddlebag
Dragonflies grab mosquitoes and eat them while flying!
Common green darner
Ebony jewelwing
Zoom In
A dragonfly's eye is made up of thousands of tiny lenses that help it see all around.

Spiders

I SEE IT!

SOME SPIDERS ARE THE SIZE OF A PINHEAD AND OTHERS ARE BIGGER THAN YOUR HAND! These hunters have eight legs but no antennae. Many use fangs to inject venom (poison) into their prey. Spiders make thin, sticky silk that they use to trap prey, build webs, and keep their eggs safe.

Yellow garden spider

Spiders often eat their old silk before spinning new webs.

Can you find some **SPIDERS**?

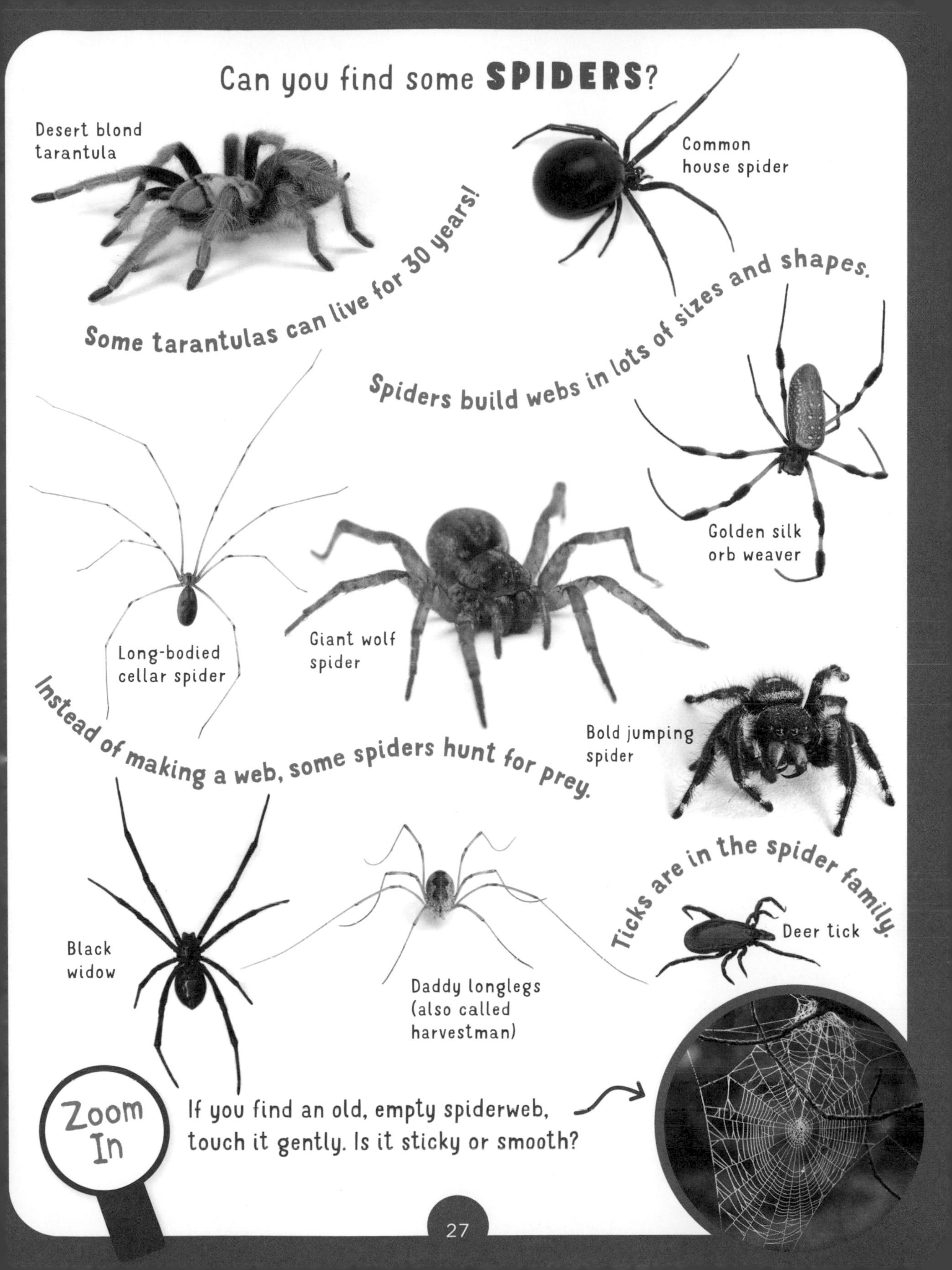

Zoom In

If you find an old, empty spiderweb, touch it gently. Is it sticky or smooth?

WEAVE A WEB

Baby spiders know how to build a web right when they hatch! Different kinds of spiders make different-shaped webs. So can you!

Gather a pile of cotton string, twine, or colorful yarn. Think about where you want to build your web. How big will it be? What color and shape?

MAKE AN ORB WEB

1 Grab a thin branch and bend it into a circle. Tie the ends together with a short piece of string.

2 Wrap string around the outside of your circle. Crisscross it back and forth until it looks like a spiderweb!

Don't be a litterbug. Please pick up all your string when you're done.

TRY A TANGLE WEB

Find a small bush, tree stump, or group of low-hanging tree branches. Thread your string in and out and around the branches to make a big messy tangle of a web!

HOW ABOUT A SHEET WEB?

This one's super easy! Weave your string back and forth and around and around over a patch of grass until it looks like a doormat.

MAKE A GIANT WEB!

Tie one end of your yarn or string to a rock, branch, or stump. Drag your string behind as you weave between bushes and around trees. As you walk, step over your string and crawl under it until you have a super big web all around you!

Grasshoppers

GRASSHOPPERS HAVE LONG, STRONG BACK LEGS MADE FOR JUMPING. Try to sneak up on one and you'll probably see it jump away fast! Most hoppers talk to each other by rubbing their wings together to make chirping sounds. Their songs can be quite noisy.

Painted grasshopper

Some grasshoppers make a loud snapping noise with their wings when they fly!

What **HOPPING INSECTS** do you see?

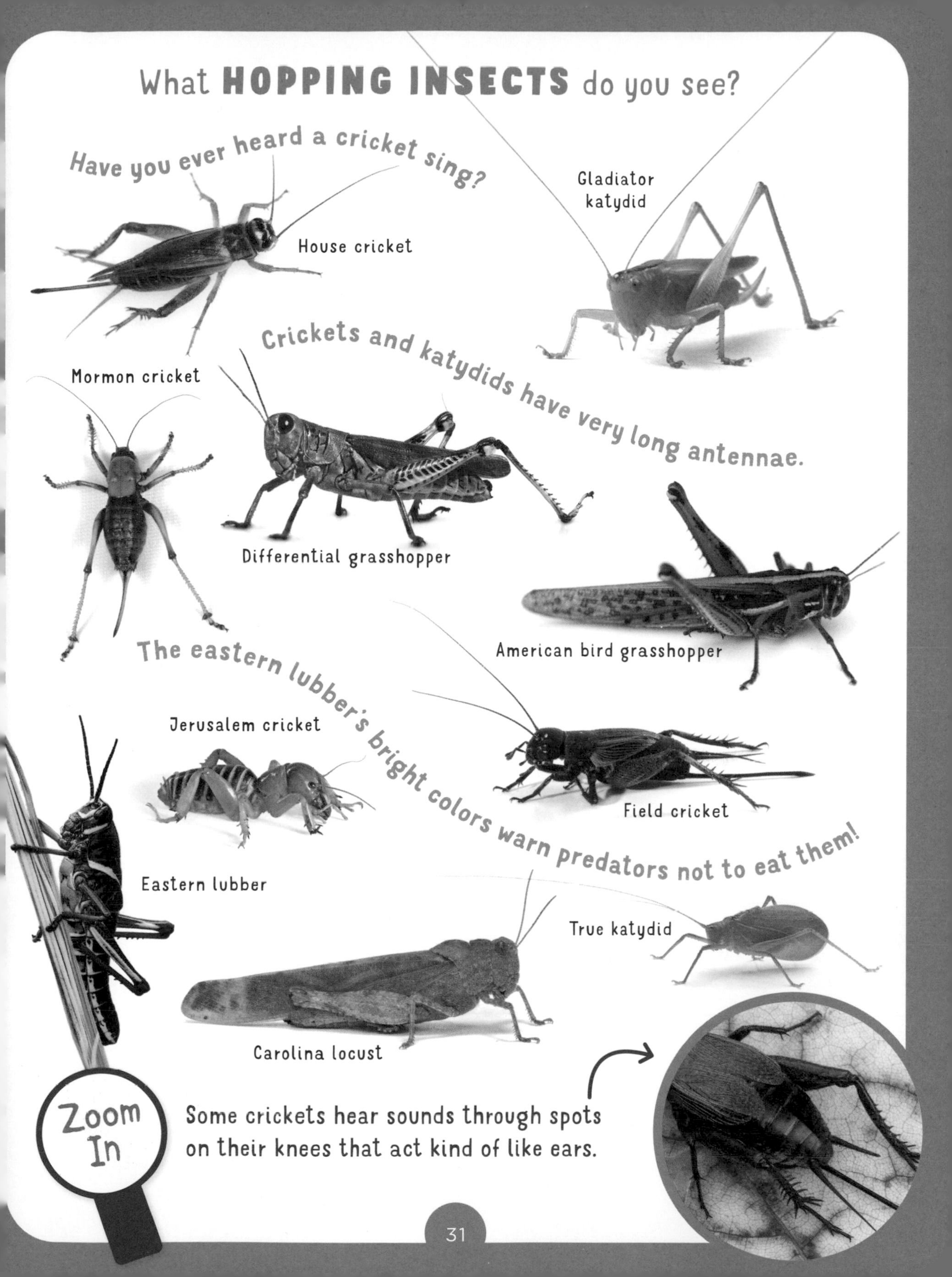

BE A BUG!

Instead of searching for bugs, pretend to be one! Try some of these simple games to get moving and have some fun.

JUMP LIKE A GRASSHOPPER!

Squat down with both feet on the ground. How high can you jump? Now see how far forward you can go!

MANTIS SPEED SNATCH

Mantises are super speedy hunters. They grab their prey so fast you can't see their legs move. You can try too!

1. Ask a friend or two to play this game. Find a pinecone, small stick, leaf, or smooth stone to be your "prey."

2. Sit on the ground facing each other or in a circle. Put the prey in the middle.

3. Stay very still until one player yells "snatch!" Now try to grab the prey!

Who's the fastest hunter?

KEEP MOVING!

Get a group of friends together
to act out these bug behaviors.

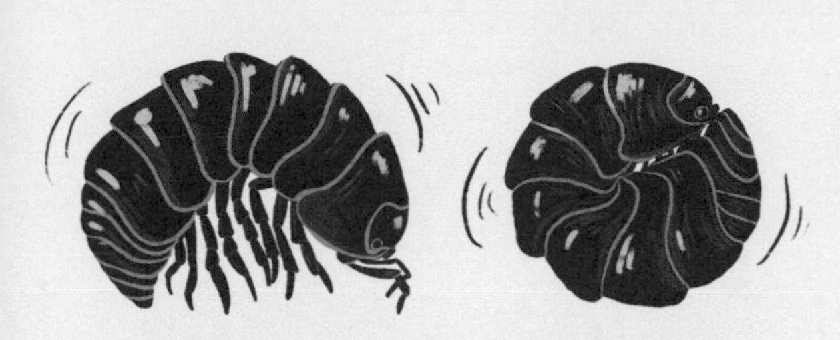

Roll up like a pill bug.

Crawl like a beetle.

Flutter like a butterfly.

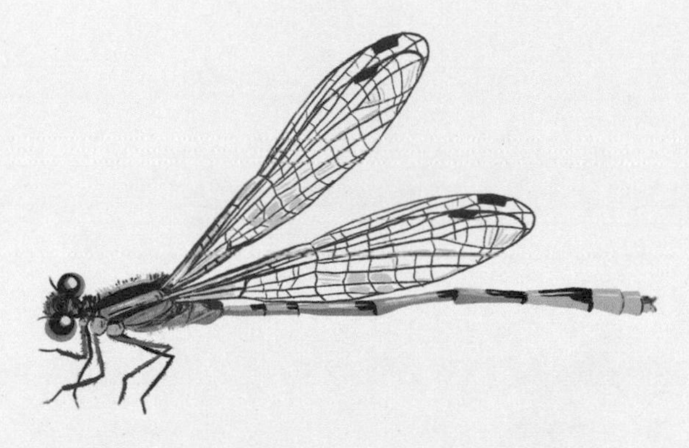

Zoom like a damselfly.

Munch like a caterpillar.

Buzz like a bee.

Flies

I SEE IT!

TRUE FLIES HAVE ONLY ONE PAIR OF WINGS, BUT THEY SURE PUT THEM TO GOOD USE. They can zigzag all around and even fly backward. A horse fly can fly as fast a speeding car! Most have a mouth like a sponge, and they use it to soak up soggy food like fruit, rotting meat, and even animal poop.

House fly

House flies can walk on walls and ceilings with special sticky pads and claws on their feet!

What other **BUZZY FLIERS** do you see around you?

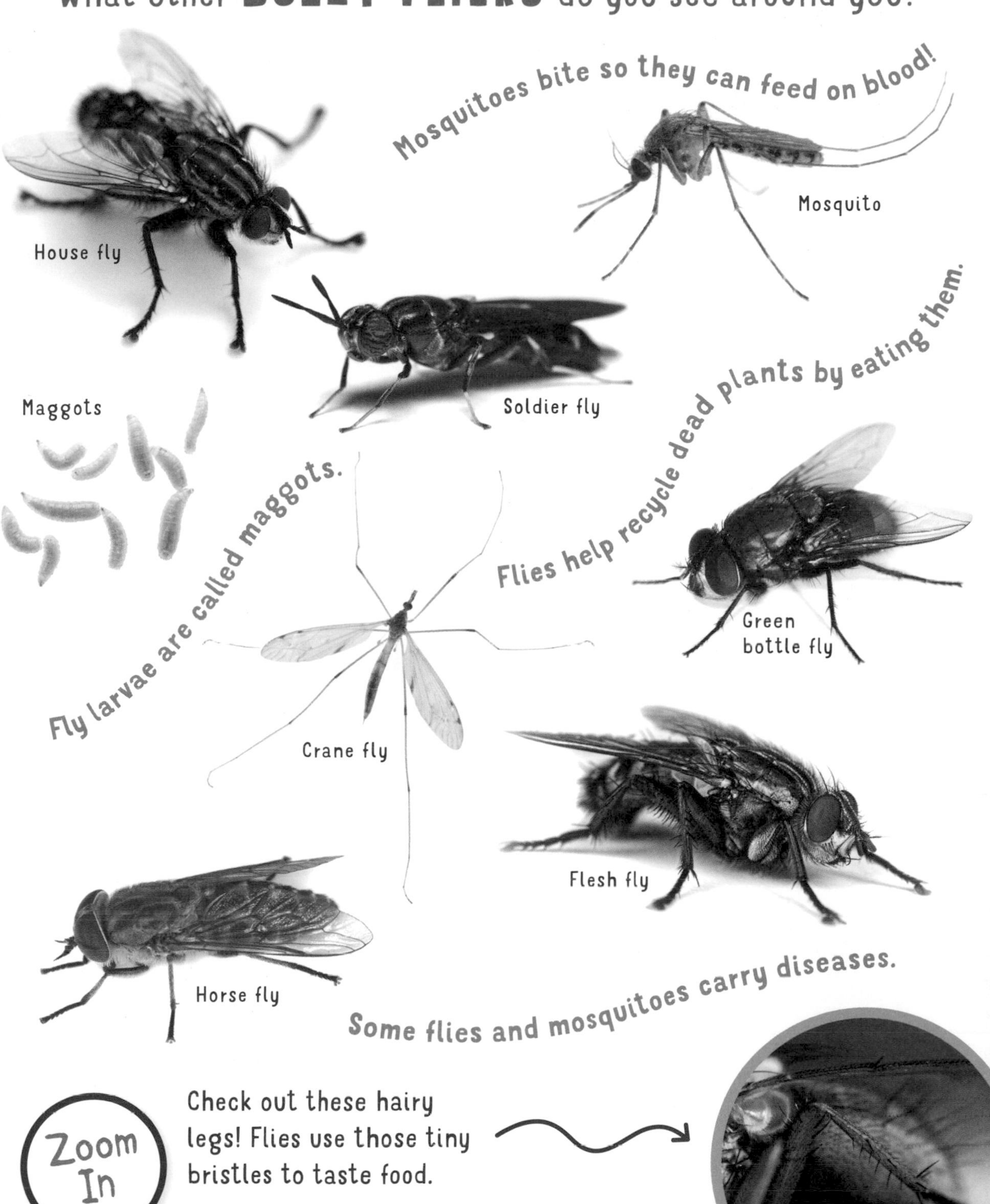

Zoom In

Check out these hairy legs! Flies use those tiny bristles to taste food.

BUILD A NATURE BUG

Want to find a one-of-a-kind bug? Make your own cool critter using stuff you find in nature.

1. Think about an insect you want to make. What does it look like? Is it colorful or camouflaged? Does it hop, fly, or crawl? Does it sting or bite, or is it harmless?

2. Look around for natural objects to use. Stones and pinecones make good heads, thoraxes, and abdomens. Colorful leaves, seedpods, and flower petals make nice wings. Try small twigs or pine needles for thin legs and antennae, and seeds or berries for eyes!

3. Arrange your materials on the ground, piecing them together however you like to design your bug buddy.

4. Now talk about it! Where does your bug live and how does it move? What does it eat? Is it noisy or silent? Does it live with family in a colony or does it wander alone?

Don't forget to give
your bug a name!

Praying Mantises

I SEE IT!

YOU'LL NEED SHARP EYES TO SPOT A BRIGHT GREEN MANTIS ON A FRESH GREEN LEAF. These insects sit very still and hide in plain sight by blending in with surrounding plants. When a tasty grasshopper or fly gets close enough — SNAP! The mantis grabs its prey fast as lightning!

While resting, a praying mantis folds its dangerous front legs in front of its body.

What other **UNUSUAL INSECTS** can you find outside?

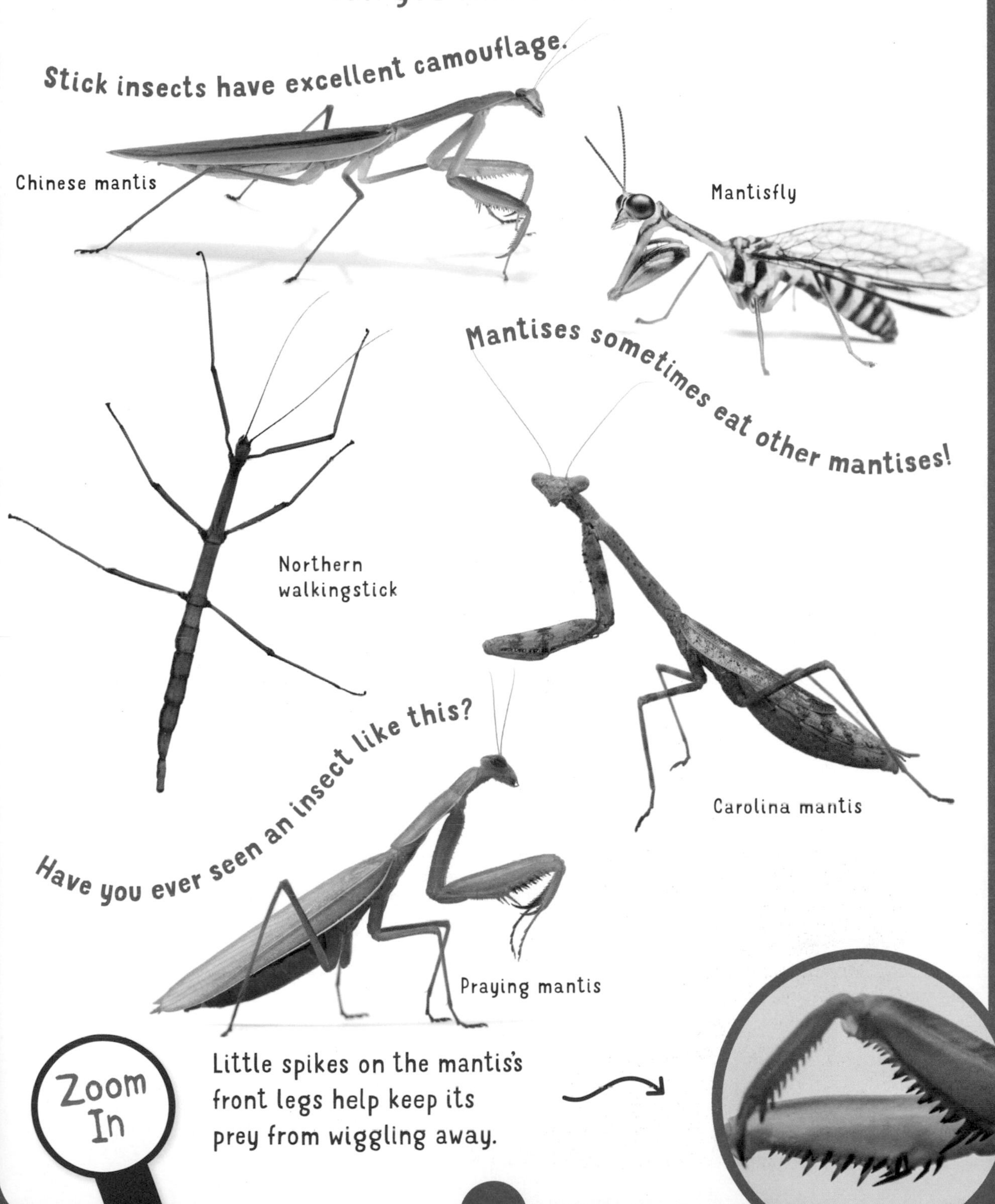

Zoom In

Little spikes on the mantis's front legs help keep its prey from wiggling away.

HIDE & SEEK

Lots of animals eat bugs, so the little guys have gotten really good at hiding to avoid being eaten. With their special colors, patterns, and amazing shapes, some bugs can almost disappear from view. Check it out!

CAMOUFLAGE

Many insects use camouflage coloring to blend in with the plants and rocks around them. They might look like a green leaf or a rough patch of tree bark, and can be very hard to find.

Some hunting bugs like mantises also use camouflage to hide and catch their prey by surprise. Can you see the bugs hidden in these pictures?

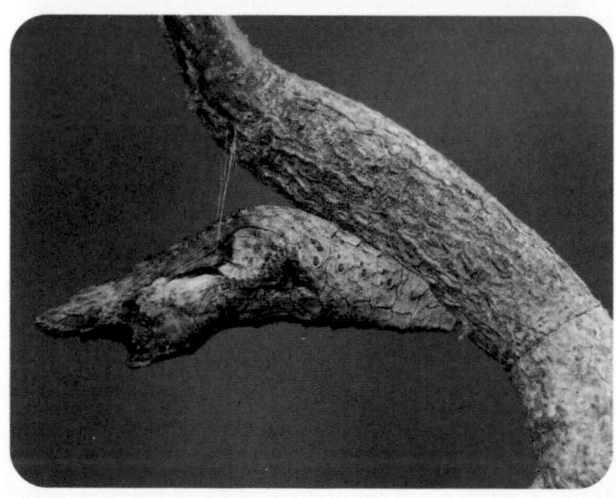

COPYCAT

Bugs can also fool predators by pretending to be other dangerous critters. Some harmless flies look like bees or wasps that might sting you. Some caterpillars and beetles have spots that look like big, scary eyes.

Spicebush swallowtail caterpillar

Eyed click beetle

Hoverfly

Worms

I SEE IT!

YOU MAY THINK THAT WORMS ARE BUGS, BUT THEY AREN'T. They don't have the legs, eyes, antennae, or hard shells of an insect. But like many bugs, worms help recycle dead plants by eating rotting leaves that have fallen to the ground. They live in underground tunnels that help keep the dirt healthy.

Earthworm

Earthworms have no eyes or ears. They sense movement through their skin.

What other **WIGGLY CRAWLERS** can you find?

Slugs and snails have sensitive horns that they pull back into their head when threatened.

HOW BUGS HELP

Bugs may be small, but they are super important. Here are some of the ways they help the planet.

GROW FOOD

Some insects pollinate flowers and help plants grow. Without the help of bees, you couldn't drink apple juice, eat strawberries, or spread honey on your toast. Other insects tunnel underground and keep the dirt healthy so roots can grow.

EAT PEST BUGS

Spiders eat lots of disease-carrying insects like mosquitoes and flies before they can bite people and other animals.

CLEAN UP

Lots of insects feed on dead plants and animals. Without their help, there would be a lot of stinky rotting stuff all over the ground.

MAKE DINNER

Many birds, bats, fish, and other animals depend on bugs for food. And guess what? Lots of people around the world eat bugs like grasshoppers, crickets, and grubs every day!

HOW BUGS HARM

A few bugs cause a lot of damage to humans and other animals. Some bugs are helpful in some ways and harmful in others.

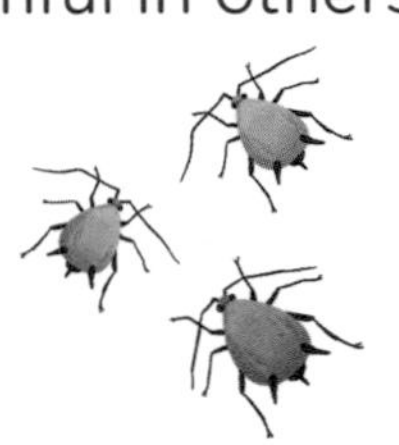

KILL PLANTS

Aphids and some caterpillars eat garden plants, and locust and grasshopper swarms can destroy entire farm crops in hours.

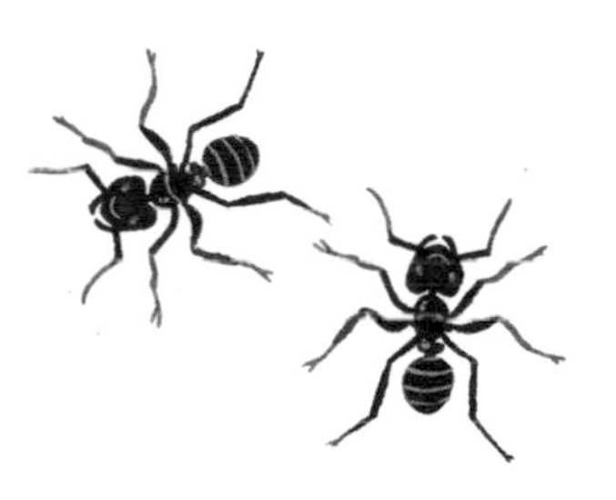

DAMAGE BUILDINGS

Carpenter ants and termites eat and burrow into wood. If there are a lot of them, they can make holes in walls and furniture.

CAUSE PAIN

Although they are small, some bugs have big ways of defending themselves! Wasps have painful stings, flies can bite hard, and some spiders and scorpions can really hurt you with their venomous bites and stings.

SPREAD DISEASE

Mosquitoes, flies, ticks, and other insects sometimes spread diseases that can make people very sick.

Water Bugs

I SEE IT!

THESE INSECTS LIVE IN SHALLOW WATER. Some, like dragonflies and caddis flies, only live in the water while they are larvae, before they transform into flying adults. Diving beetles, water boatmen, and backswimmers stay in the water their entire lives.

Water strider

Instead of swimming, many water bugs skim along the surface of the water.

Can you find an insect that lives **IN THE WATER**?

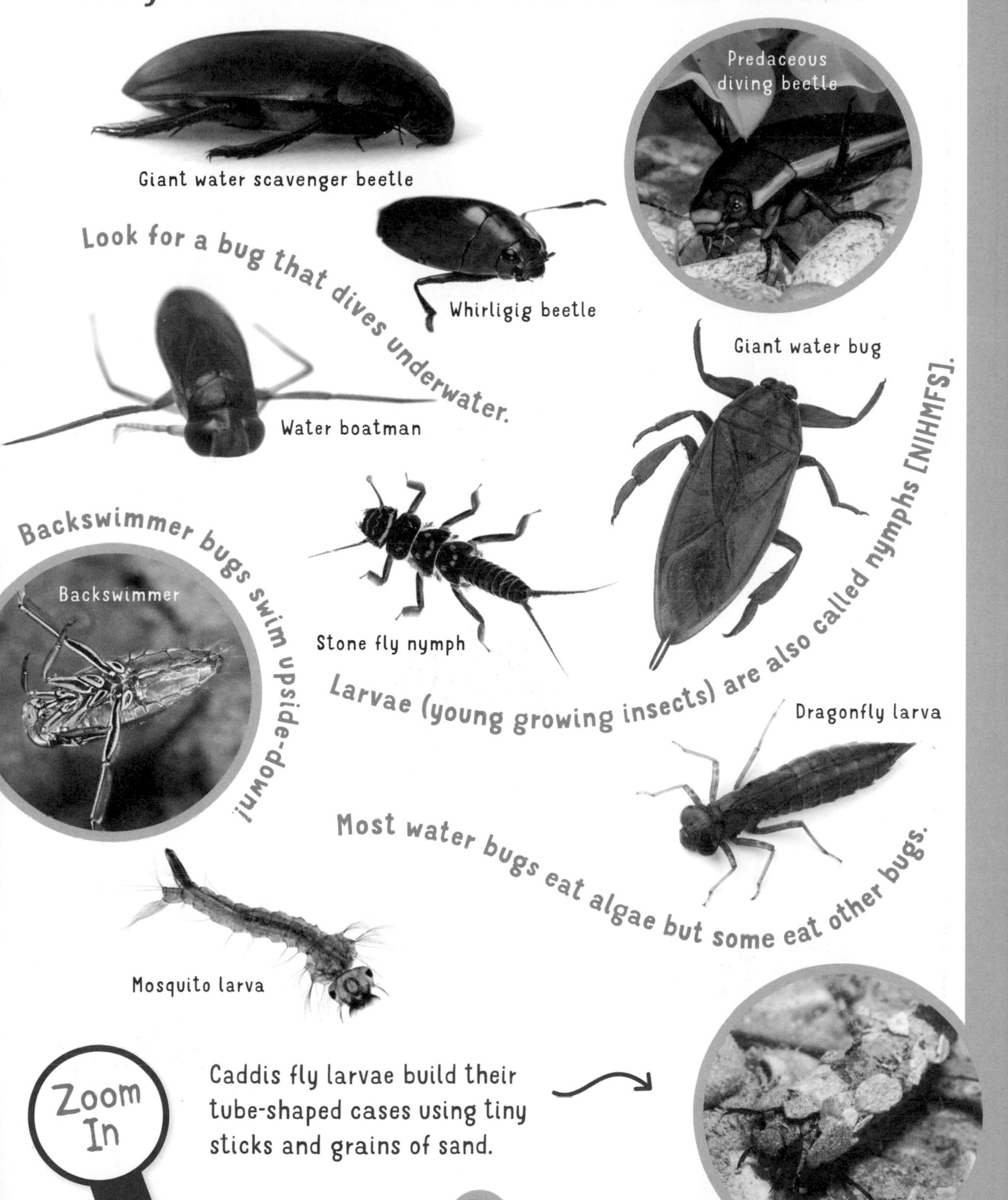

Caddis fly larvae build their tube-shaped cases using tiny sticks and grains of sand.

BUG FOOD

Different insects have very different tastes in food. Do you see any of these bug foods on your nature walk? Don't touch them. Just check off the box!

MY DAY OF BUG HUNTING

Match up stickers to what you saw on your bug hunt.

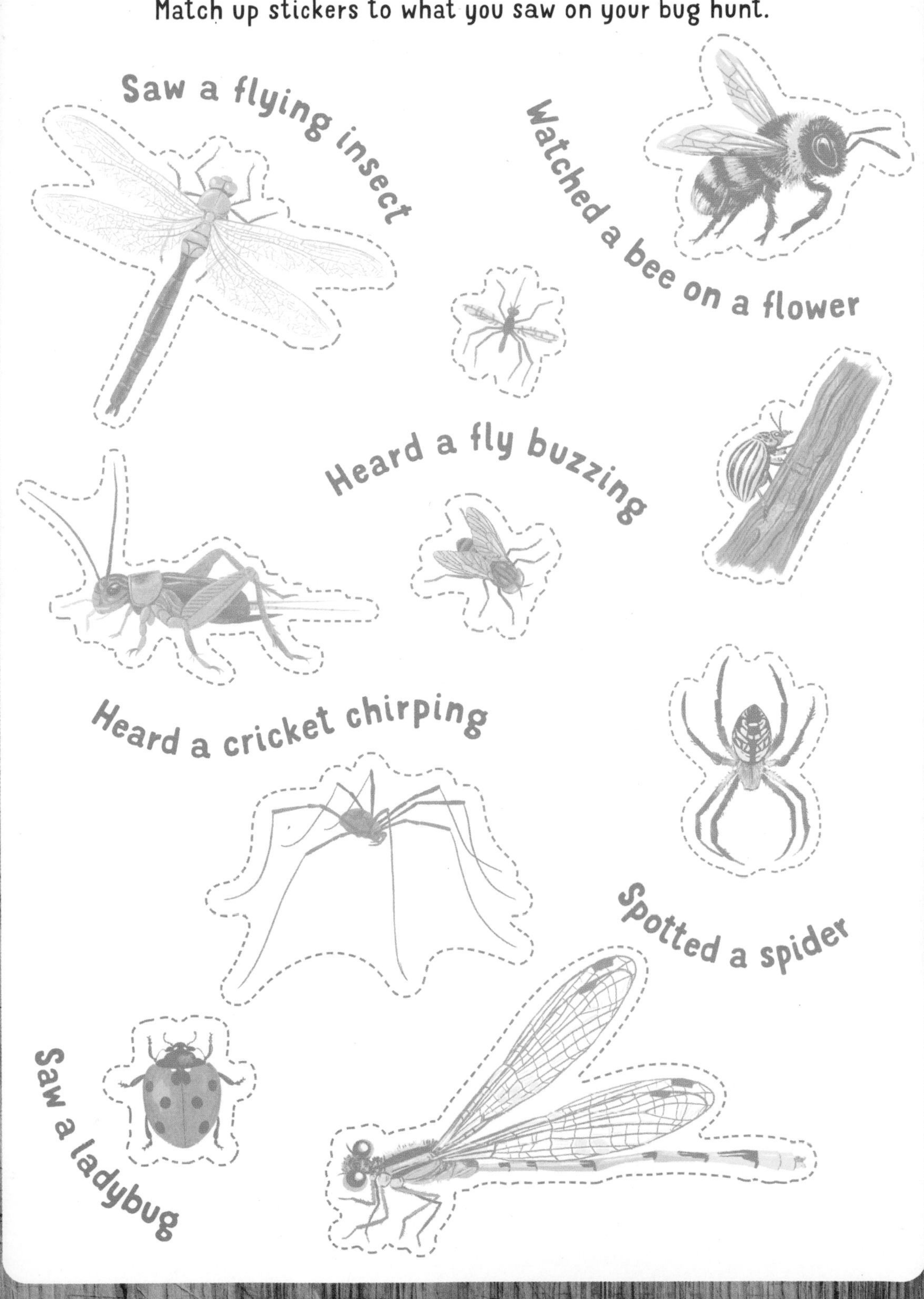

MY DAY OF BUG HUNTING

Match up stickers to what you saw on your bug hunt.

I saw these bugs:

Draw a picture or add some stickers of the bugs you saw on your bug hunt. Write their names. What were they doing when you saw them?

MY BUG LOG

Keep a list of what you find on your bug-hunting adventures.

Bug: ______________________ **Date:** __________

Where was it? ______________________ **Color:** __________

What was it doing? ______________________

Bug: ______________________ **Date:** __________

Where was it? ______________________ **Color:** __________

What was it doing? ______________________

Bug: ______________________ **Date:** __________

Where was it? ______________________ **Color:** __________

What was it doing? ______________________

Bug: ______________________ **Date:** __________

Where was it? ______________________ **Color:** __________

What was it doing? ______________________

Bug: ______________________ **Date:** __________

Where was it? ______________________ **Color:** __________

What was it doing? ______________________

Bug: ______________________ **Date:** __________

Where was it? ______________________ **Color:** __________

What was it doing? ______________________

The mission of Storey Publishing is to serve our customers by publishing practical information that encourages personal independence in harmony with the environment.

Text by Kathleen Yale
Edited by Deanna F. Cook and Lisa H. Hiley
Art direction and book design by Michaela Jebb
Text production by Jennifer Jepson Smith

Illustrations by © Oana Befort
Photography by Mars Vilaubi © Storey Publishing, 15, 18, 28, 29, 36, 37
Additional stock photography by © Aflo Co., Ltd./Alamy Stock Photo, 47 (Giant water bug); © Alex Wild/alexanderwild.com, 17 (Honeypot ant, Kidnapper ant); © aleksandarfilip/stock.adobe.com, 43 (Garden snail); © Amelia/stock.adobe.com, 23 (chewed leaves); © Amy Olson/Alamy Stock Photo, 25 (Blue-tailed damselfly); © anake/stock.adobe.com, 22; © ariadna126/stock.adobe.com, 23 (cocoon or chrysalis); © Arterr Picture Library/Alamy Stock Photo, 35 b.r.; © Auk Archive/Alamy Stock Photo, 47 (Backswimmer); © Ayupov Evgeniy/stock.adobe.com, 13 (Yellowjacket); © Babar760/Fotosearch LBRF/agefotostock, 27 (Deer tick); © benz190/iStock.com, 3 (cork board); © Bill Gozansky/Alamy Stock Photo, 24; © blickwinkel/Alamy Stock Photo, 27 (Desert blond tarantula), 43 (Megascolecidae worm); © Brad Sharp/agefotostock, 27 (Bold jumping spider); © Brian Kushner/Alamy Stock Photo, 25 (Blue dasher); © Buddy Mays/Alamy Stock Photo, 2 & 31 (Jerusalem cricket); © Buitem-Beeld/Alamy Stock Photo, 42; © Cavan/stock.adobe.com, 25 (Twelve-spot skipper); © chas53/stock.adobe.com, 5 (Isabella tiger moth); © Christian Buch/stock.adobe.com, 34; © Christina/stock.adobe.com, 23 (egg sac); © Cristina Lichti/Alamy Stock Photo, 39 (Northern walkingstick); © Christina Rollo/Alamy Stock Photo, 25 (Ebony jewelwing); © ConstantinCornel/iStock.com, 25 b.r.; © Danita Delimont/Alamy Stock Photo, 25 (Eastern pondhawk), 40 t.l. & b.l.; © defun/stock.adobe.com, 47 (Whirligig beetle); © Design Pics Inc/Alamy Stock Photo, 39 (Chinese mantid); © Diana Meister/Alamy Stock Photo, 12; © dimakp/stock.adobe.com, 21 (Potato beetle); © Dragi52/iStock.com, 9 b.r.; © Ekaterina Bogomolova/EyeEm/Getty Images, 27 b.r.; © eleonimages/stock.adobe.com, 39 b.r.; © elharo/stock.adobe.com, 17 (Red imported fire ant), 21 (Big dipper firefly); © Elliotte Rusty Harold/Shutterstock.com, 17 (Pavement ant); © Eric Isselée/stock.adobe.com, 27 (Harvestman), 43 (Leopard slug); © Floydine/stock.adobe.com, pull-out (background); © Gay Bumgarner/Alamy Stock Photo, 41 t.l.; © Gerry/stock.adobe.com, 21 b.r.; © grandaded/stock.adobe.com, 35 (Flesh fly); © Grant Heilman Photography/Alamy Stock Photo, 13 b.r.; © guy/stock.adobe.com, 13 (European honeybee); © hakoar/stock.adobe.L8com, 2 & 30, 2 & 31 (Eastern lubber), 25 (Red saddlebag); © HeitiPaves/iStock.com, 9 (Painted lady butterfly); © Henrik Larsson/stock.adobe.com, 35 (Horse fly), 47 (Water boatman); © hhelene/stock.adobe.com, 35 (Green bottle fly, House fly); © Holly Guerrio/stock.adobe.com, 23 (galls in sticks); © hsagencia/stock.adobe.com, 13 (European hornet); © Ian Butler Photography/Alamy Stock Photo, 9 (Green hairstreak butterfly); © James L Davidson/stock.adobe.com, 27 (Black widow); © Jaroslav Machacek/stock.adobe.com, 43 b.r.; © Jason Ondreicka/Alamy Stock Photo, 47 (Predaceous diving beetle); © Jim Larkin/Shutterstock.com, 2 & 31 (Differential grasshopper); © John Abbott/NPL/Minden Pictures, 17 (Red harvester ant); © Juniors Bildarchiv GmbH/Alamy Stock Photo, 2 & 31 (House cricket); © kertlis/iStock.com, pull-out (meadow); © Kevin/stock.adobe.com, 16; © kirin_photo/iStock.com, 19 b.; © KPixMining/stock.adobe.com, 5 b.r.; © leekris/iStock.com, 8, 9 (Viceroy butterfly); © Leena Robinson/Alamy Stock Photo, 26; © Leigh Prather/Shutterstock.com, 5 (Luna moth); © Leon Werdinger/Alamy Stock Photo, 43 (Banana slug); © Life on white/Alamy Stock Photo, 27 (Common house spider), 39 (Praying mantis); © MarkMirror/iStock.com, 9 (Mourning cloak butterfly); © Martha Marks/stock.adobe.com, 5 (Mourning cloak); © Martin Shields/Alamy Stock Photo, 46; © Matt Jeppson/Shutterstock.com, 5 (Cecropia moth); © Melinda Fawver/stock.adobe.com, 2 & 31 (American bird grasshopper, True katydid), 5 (© Melinda Fawver/Shutterstock.com, 25 (common green darner); Milkweed tiger moth, Tobacco hornworm moth), 13 (Bald-faced hornet), 21 (Click beetle larva, Green June bug, Rainbow scarab dung beetle), 27 (Long-bodied cellar spider), 47 (Giant water scavenger beetle); © Michael Siluk/Alamy Stock Photo, 2 & 31 (Field cricket); © Mindy Fawver/Alamy Stock Photo, 13 (Paper wasp); © MYN/Cly Bolt/npl/Minden Pictures, 13 (Metallic green sweat bee), 21 (Eastern Hercules beetle); © MYN/Pierre Esccoubas/npl/Minden Pictures, 13 (Black-and-yellow mud dauber); © National Geographic Image Collection/Alamy Stock Photo, 2 & 31 (Gladiator katydid), 13 (Bumblebee), 21 (Ten-lined June beetle), 25 (Familiar bluet); © Nikolay N. Antonov/stock.adobe.com, 35 (Maggots); © Pat Canova/Alamy Stock Photo, 23 (ant mound); © paule858/iStock.com, 17 b.r.; © Paweł Burgiel/stock.adobe.com, 47 (Mosquito larva); © peerayut aoudsuk/EyeEm/stock.adobe.com, 23 (spider web); © peter/stock.adobe.com, 17 (Big-headed ant); © Petlin Dmitry/Alamy Stock Photo, 21 (Giant stag beetle); © phichak/stock.adobe.com, 35 (Soldier fly); © pimmimemom/stock.adobe.com, 4; © Rachel Kolokoff Hopper/Alamy Stock Photo, 41 b.; © Rolf Nussbaumer Photography/Alamy Stock Photo, 38; © Rostislav/stock.adobe.com, 47 b.r.; © Saddako/iStock.com, 21 (Eyed click beetle); © Samuel Ray/Alamy Stock Photo, 39 (Carolina mantid); sander boot/Unsplash, 20; © Savany/iStock.com, 3 (push pins); © sbeagle/iStock.com, 19 t.; © Science History Images/Alamy Stock Photo, 23 (beetle holes in bark); © Shawn Coates/iStock.com, 17 (Black carpenter ant); © Skip Moody/Dembinsky Photo Associates/Alamy/Alamy Stock Photo, 2 & 31 b.r., 40 r., 41 t.r.; © Stephanie Frey/stock.adobe.com, 9 (Spicebush swallowtail butterfly); © SteveByland/iStock.com, 9 (Hummingbird moth); © SunnyS/stock.adobe.com, 5 (Monarch caterpillar), 27 (Golden silk orb weaver); © The Natural History Museum/Alamy Stock Photo, 9 (Cloudless sulfur butterfly); © thithawat/stock.adobe.com, 35 (Mosquito); © Thomas J. Peterson/Alamy Stock Photo, 17 (Argentine ant); © thongsan/stock.adobe.com, 23 (empty exoskeleton); © Todd Bannor/Alamy Stock Photo, 5 (Giant swallowtail butterfly); © Tonia/stock.adobe.com, 23 (wasp nest); © triggermouse/iStock.com, 2 & 31 (Mormon cricket); © troutnut/stock.adobe.com, 47 (Stonefly nymph); © Valentina R./stock.adobe.com, 43 (Lumbricidae worm); © Vitalii Hulai/stock.adobe.com, 47 (Dragonfly larva); © Vonkaral/iStock.com, 9 (Buckeye butterfly); © VStock/Alamy Stock Photo, 27 (Giant wolf spider), 35 (Crane fly); Will Brown/CC BY 2.0/Wikimedia, 2 & 31 (Carolina locust); © wizardofwonders/stock.adobe.com, 9 (White-lined sphinx moth); © Zety Akhzar/Shutterstock.com, 39 (Mantidfly)

Storey Publishing
210 MASS MoCA Way
North Adams, MA 01247
storey.com

Printed in China by R.R. Donnelley
10 9 8 7 6 5 4 3 2 1

Library of Congress Cataloging-in-Publication Data on file